Britain's Wildlife

Britain's Wildlife

Brian Grimes
Photographs by John Markham
Foreword by Eric Hosking

Collins
Glasgow and London

First published 1974

Published by
William Collins Sons and Company Limited,
Glasgow and London

© 1974 William Collins Sons and Company Limited

Printed in Great Britain

ISBN   0 00 106144 5

Cover: Harvest Mouse; Robin
Front endpapers: Pine Marten (left); Yellow-hammer (right)
Title page: Coots
Back endpapers: Ptarmigan (left); Siskin (right)

Contents

Foreword	10
Introduction	12
Birds	17
Mammals	57
Reptiles	79
Worms, slugs and leeches	87
Insects	91
Plants	103
Index	124

Foreword

John Markham was my friend and colleague for many years. We had much in common, being both north-Londoners with interests in photography and wildlife. In my capacity as photographic editor of the Collins' *New Naturalist* series, I commissioned him to photograph many varying subjects including plants, mammals, birds, reptiles, insects and sometimes scenery. The results were photographic masterpieces.

He was, in the opinion of myself and many of my colleagues, a most versatile British wildlife photographer. His knowledge of natural history was as extensive as his mastery of the mechanics of photography and photographic apparatus.

He was a professional whose work has been reproduced in countless publications in all parts of the world. However, few people knew the real John Markham for he usually shunned most meetings, preferring to carry out his photography alone. Only once in his whole career did he spend a summer season with another photographer. This was in Norfolk in 1955 when he and Walter E. Higham photographed little owls.

Because he worked so much on his own it is difficult to say much about the man. He told me his interest in birds began at the age of three; as soon as he was able to read, natural history books were his first choice. In particular the books of Oliver Pike and the Kearton brothers sparked off his interest. It was a red-letter day when he met that great nature photographer of the twenties, Oliver Pike, who promptly advised him to get a camera. John Markham bought a simple box camera and — with the aid of a tripod and such home-made gadgets as a spectacle glass over the lens to reduce the focal length — he was able to produce acceptable photographs of nests and eggs, and eventually of birds. His sphere of attention spread to the insects eaten by the birds and, in turn, to the plants on which the insects lived. Always his interests were with the common rather than the rare and with behaviour and basic ecology rather than photographic portraiture.

He would spend weeks, even months, on one particular species as, for example, the dotterel in the Cairngorm mountains in Scotland. Year after year he climbed to its breeding grounds and often stayed out all night; in June, of course, it never becomes dark at that latitude. Another season he devoted his time to the photography of the woodcock — an extremely difficult bird to photograph due to its effective camouflage. While another whole nesting period was spent on the nightjar. A particular challenge was the grasshopper warbler which he wanted to portray singing; he gave practically the whole of the 1962 season to watching this species in Anglesey, finding its favourite song perches and establishing a method of stalking the singing bird until close enough to photograph it.

He would sit by his camera for days on end until he obtained the photograph he wanted. He spent long hours in a hide baited with food and this resulted in photographs of a variety of birds; his photograph of the willow tit is, in my opinion, the best of them. He was also an opportunist. One day in April 1951, when driving along a narrow country lane in Devon, he stopped to eat his sandwiches. Almost at once a chaffinch came down to look for crumbs and then flew up to a perch on a spray of apple blossom; the resulting photograph was considered to be one of the best ever obtained of this species.

Many of his photographs are of birds at or near their nests and, here again, John Markham showed his outstanding skill. His photograph of the stone curlew is well worth mentioning because the female is clearly oblivious of the photographer. John Markham always exercised a strict discipline in his behaviour towards the subject matter and would not disturb the bird more than was absolutely necessary. Like most bird-photographers, he explored the use of roll film instead of the traditional plate camera; his photograph of the wood warbler at its nest surrounded by ferns and beech leaves is a good example of his success in this field.

His work in its inimitable style will for some time to come be an inspiration to wildlife photographers searching for perfection.

Eric Hosking

Willow Tit

Garden Warbler at night

Dotterel

Introduction

John Markham was only three years old when his mother pointed out a starling singing on a chimney pot. This incident was to have a profound effect on the young boy. From that moment his interest in nature, and in particular bird life, had begun. This interest and his method of expressing his appreciation — photography — were to bring him world-wide acclaim. John Markham died at the age of 70 on September 28, 1972. This book is a tribute to the man who went on, from humble beginnings, to become a Fellow of the Royal Photographic Society, a Fellow of the Zoological Society and the country's finest all-round natural history photographer.

In another tribute to John Markham, in *The Photographic Journal* last year, his friend and colleague Eric Hosking, FRPS, commented: 'Those who knew John will miss his personality and dedicated approach to this work, but his photography will live on for many years and may we hope that it will inspire some young enthusiasts to explore and record with their cameras the wonderful field of nature, just as he in his turn was made aware of the beauties around him'.

John Markham was born in Wood Green, North London on November 22, 1902. Even before he started school the seeds of interest in nature had been sown. They were to grow and flourish until his death. As a schoolboy, his passion for fauna and flora took him on regular trips to the countryside, usually by cycle to collect specimens. Then followed hours of careful identification from library and reference books. Most of his plant specimens were pressed in scrap books. In fact, so expert — and methodical — did the young Markham become, that his knowledge of nature soon outstripped that of his schoolmaster. And it was while he was still young that another chance incident, one he was to remember all his life, helped to stimulate and reinforce his interest in bird watching and photography.

John was given a copy of Oliver Pike's book *Hillside Rock and Dale*. From then on Oliver Pike became a hero to the young boy and in fact the two were to meet; but not at first under the best of circumstances.

One day John was out on one of his nature tours when he was accused of taking the eggs from a nest which Oliver Pike was photographing. John was given a stern lecture and told the best thing he could do when he was old enough was to buy himself a camera.

After using his sister's camera to attempt to photograph birds in the garden John's enthusiasm grew. He made a journey to the old Caledonian Market in North London where he bought a second-hand box Brownie camera. With a few home-made gadgets, a tripod and his box Brownie, the young Markham set off with more enthusiasm than skill to take his first pictures of birds at their nests. One of these photographs was of a brooding robin nesting in ivy on a yew tree. The result, he confided later, looked more like a wild duck than a robin but it was a beginning of which he was proud.

By the time he was 14, John and a friend with similar interests were making regular nature excursions to nearby estates. His family would often find after his expeditions that his bedroom had become a temporary home for frogs, toads, voles and other animals he wished to study more closely. John had already learned to identify most of the mammals, and was now more interested in trying to photograph them. But, as with the robin, he was not satisfied with his first results.

The interest in the common as opposed to the rare, and in behaviour rather than just portraits of his subject, was to stay with him all his life. Eric Hosking described it like this: 'Sometimes he would spend weeks, even months, on one particular species trying to record with his camera almost every detail of its life. During the 1950s he made the dotterel his target and year after year climbed to its breeding grounds on the tops of some of Scotland's highest mountains to obtain a fine series of photographs. Another year he worked almost exclusively on the woodcock and he was most disappointed not to obtain a photograph of an adult carrying its young — he was convinced that it did.'

Even at this age he was developing the attitudes of the perfectionist. The mediocre and good were not acceptable — only perfection really satisfied him. This early passion for perfection was reflected in later years when, with his knowledge and artistic eye, he would create beauty out of the commonplace. And so the young Markham set out to read all the books he could on photography. His knowledge grew — and he was entirely self-taught.

After leaving school he went to work in various offices; but his heart was not in the work. His mind would often wander to the countryside and his beloved birds and flowers. Work was nothing more to him than a necessity between his nature trips. Almost all his spare moments were spent, usually alone, in the country, learning about natural history and taking photographs. At this time he spent a lot of time at the sewage farm near the Great Northern Cemetery and in the cemetery itself which, in parts, was very wild.

It was on the sewage farm that he managed to get a photograph of a lapwing, which was to be one of the first photographs he submitted to the Royal Society. It was, to his great disappointment, rejected — but was accepted the following year.

John had discovered that, by clipping an ordinary spectacle glass to the lens of his Brownie, he could work at closer range. But for his serious photography he graduated to a $\frac{1}{4}$-plate field camera fitted with a twelve-inch Dallon tele-photo lens and an LUC shutter. John was still living in London and did not like the idea of using hides; he thought they would attract attention, and that the birds would desert their nests. Instead he used remote control — a long length of rubber tubing with a bulb at the end — to take close-up photographs of small birds feeding their young. After each exposure he would wait until the parent bird had flown away before changing the plate. He did, of course, use hides but usually in the countryside. One of his most successful early photographs was of the red-throated diver. John travelled to an island off the Scottish coast where he soon located a nest. He erected a hide about 50 yards away and, each evening and morning, returned to place it a little nearer. Eventually it was only 10 feet away. This part of the operation over, he had to persuade the bird that the hide was empty. He enrolled the help of a local boy who drove him up the rough track and then left. Before the cart was out of earshot one of the divers had returned to its nest. The picture John took was to appear a few years later in *Masterpieces of Bird Photography* published by Collins. John's knowledge of birds — and insects and plants — grew and he often set himself on a course of intensive search for detail. He would sometimes spend months painstakingly obtaining slides on a particular bird species and would not usually be satisfied until he had achieved his objective.

In the 1930s John was assisting his uncle in his decorating business, but nature and photography were still his passions. Although still an amateur, the name of John Markham was becoming better known. His pictures and articles were appearing in publications such as *Country Life, London Calling, The Field, The Sphere, Illustrated London News* and *Gun & Game*. Also at this time he accepted many invitations to judge nature photographic exhibitions. He supplied the photographs for one *Brooke Bond Tea* picture card album.

Towards the end of 1941 John was invited by his friend, Eric Hosking, to produce a series of high quality colour photographs for The *New Naturalist* series of books published by Collins in conjunction with Adprint. John was now an air-raid warden and often worked 48 hours on, 48 hours off. Eric Hosking later wrote of this time: 'I supplied him (John) with colour film and told him what subjects I wanted. He always came back with first class results. It did not matter whether it was a scenic shot of Beachy Head, a carpet of bluebells, a spider spinning its web, fungi growing on the floor of a wood, a fish in a tank, a mouse emerging from its hole or a bird feeding its young, the colour transparency was as near to perfection as it was possible to get.'

It was after the war that John decided that his future was in photography and he became a full-time professional photographer. The man who had dedicated so much of his spare time to his hobby could now make it his whole life. He had become a professional in every sense of the word. A long and distinguished career was to follow.

In 1943 John joined the Royal Photographic Society, becoming an Associate in 1946 and a Fellow the same year. He served on the Exhibitions Committee from 1958 to 1968, and from 1953 he served almost continuously on the Fellowship and Associateship Admission Committees (Nature) Panel until only a few years ago. He was a founder member of the Nature Conservancy's Advisory Committee on Photography and was instrumental with Eric Hosking and Harold Lowes in creating the National Collection of Nature Photographs. To this collection he presented over a hundred of his best photographs. He was a member of many societies connected with natural history – the Mammal Society of Great Britain, the Royal Society for the Protection of Birds, the Zoological Photographic Club and the Nature Photographic Society. He also won numerous medals, certificates

Red-throated Diver

and other awards in various photographic exhibitions in many parts of the world.

His accolades and achievements are self-evident. He combined skill and knowledge of the mechanics of photography with the eye of an artist. His photographs represented hours, sometimes weeks, of patient work. He had, as all great natural history photographers must have, great patience. Patience to wait and wait for the right moment – the moment that divides the good from the brilliant. On one occasion he saw a wood mouse running under a hedge beside a wood. That particular species had never been photographed in colour in the wild. He focused his camera on a spot where he thought the mouse would appear and waited, sitting on his heels. The waiting went on and on until finally the mouse did appear. John tried to press the shutter release. Nothing happened. His hand was numbed with cold.

Yet for all his many talents he was a man who hated change. He called 35mm single lens reflex cameras 'toys' and was opposed to their use. However, he did produce a series of slides, using a single lens reflex camera, of garden birds, flowers and insects – but later continued to use the larger format cameras. He said he felt that, however good the 35mm might be, the photographs were too small to be seen unless enlarged. The introduction of high speed flash also caused him some initial concern. He was always so careful not to disturb his subject; he was convinced that the brilliance would upset any wild creature. It

Wood Mouse

was not until he was taking photographs in Norfolk with a colleague who used flash, and saw what little notice the creatures did take of it, that he began to experiment himself. The resulting pictures of animals in action were – as one had come to expect – superb. Although he did not derive pleasure from writing, when he did write about photographing plants and animals, John wrote in an easy style and with great simplicity. One such article which appeared in *The Field Annual on Nature Photography* is a fine exposition of the problems confronting the would-be natural history photographer. It describes methods in overcoming many of the difficulties and gives advice on the types of camera to use. In particular, it emphasises the importance of resisting the temptation to open up the vegetation surrounding nests in order to improve the lighting conditions. Not only will this practice cause suffering to the bird, it can cause desertion of the nest even if it contains eggs or young. It is also obvious to anyone with knowledge of birds and their nests when disturbance of the nest has taken place.

John Markham was a founder member of the Committee on Photography, which advises officials regarding the granting of licences to photographers wishing to photograph rare birds contained in the Bird Act 1967.

Invariably, John advised against the issuing of a licence if there was evidence of 'gardening' shown in the prints submitted by the photographer.

When he died last year the name of John Markham had become synonymous with perfection. But although the man is dead his work will live on as a delight and an inspiration.

Right: Linnet

Pied Flycatcher *Ficedula hypoleuca*
The male bird as shown in this photograph is very strikingly marked with black head and upper parts and white wing patches and sides of tail. The female is a brownish colour with smaller wing patches. They can be found mainly in deciduous woodlands in Wales and south-west England and sometimes in gardens. They nest in holes in trees and walls, and in nest boxes.

Chaffinch *Fringilla coelebs*
The commonest of the finches in Britain, which can be seen in a wide variety of habitats – the garden, woods, farmland, commons and hedgerows. Its undulating flight is easily recognisable. The male is attractively marked with its slate blue crown and nape, a brown-pink breast and green rump. The female is less attractively marked and is a dull olive-brown. It is widespread throughout Europe. This much-admired photograph was taken at the roadside after a car breakdown.

Coal Tit *Parus ater*
The most distinguishing feature of this tiny bird, which measures only four and a quarter inches, is the bold white patch on the nape of its neck; this white patch clearly identifies it from the willow and marsh tits. They are less bold than the blue tits. Both sexes are alike. They nest in holes in trees, banks or walls and will use nest boxes. They can be found throughout Europe and parts of Asia.

Marsh Tit *Parus palustris*
The marsh tit is very similar to the willow tit, but it has a brighter, black crown and a distinctive 'pitchew' call. It has no particular liking for marshes, but is to be found mainly in woods, hedges and thickets. It is not as common as the great and blue tits in the garden.

Young Kingfishers *Alcedo atthis*
This beautiful bird can be seen flying low over rivers and streams; the flash of its iridescent colours is a magnificent sight. It will perch on branches over-hanging water for long periods, waiting for fish. Then it plunges headlong into the water.

Nightingale *Luscinia megarhynchos*
It is famous throughout southern and western Europe for its beautiful song. It confines itself to rather wet undergrowth of woodlands and sometimes frequents parks. It is very secretive, keeping well hidden in the undergrowth.

Sand Martin *Riparia riparia*
This is the smallest European swallow, measuring only four and three-quarter inches. Its slim, streamlined form and graceful flight are distinctive. The sand martin catches its food in flight. Its nesting habits differ from those of swallows and other martins; it burrows in sand-banks and cliffs. In late summer most sand martins leave Britain to winter in Africa. Its brown breast-band and white under-parts distinguish it from the house martin.

Pied Wagtail *Motacilla alba*
This well-patterned black and white bird with its slender legs and long, wagging tail can be found throughout Europe. The female's markings are much greyer than those of the male. It can be found in gardens, farms, parks and the country. One of its favourite feeding habitats seems to be a well-cut lawn.

Garden Warbler *Sylvia borin*
The warblers are a very numerous family; They feed almost exclusively on insects. The garden warbler has few distintive features and is best identified by its song, although this can be mistaken for the song of the blackcap.

Great Tit *Parus major*
Its larger size immediately distinguishes the great tit from other tits. It is a very handsome bird with its striking black head and bib, white patch below the eyes and yellowish underparts. It appears to be less sociable than other tits, but in the winter can occasionally be seen in flocks with other tits and nuthatches.

Swift *Apus apus*
Swifts can be distinguished from swallows by their long scythe-like wings, short tails and very rapid flight. They are exclusively aerial birds and feed on insects taken on the wing. They can be very noisy and will often fly low giving out a shrill, piercing screech. Their nests are made of straw and grass cemented with saliva and are situated under eaves or tiles of buildings or in rock crevices in caves or cliffs.

Blue Tit *Parus caeruleus*
This cheeky, agile bird is one of Britain's favourite garden visitors. Its habit of taking the cream out of milk bottles causes some displeasure. This picture was taken outside the photographer's house; he would not allow anyone to stop the birds taking the cream. The bird is beautifully coloured with its light blue crown and wings. The blue tit takes readily to nesting in boxes.

Wood Warbler *Phylloscopus sibilatrix*
It can be distinguished by its broad yellow stripe above the eye, brightly contrasted yellow-green upper parts and its white belly. It has a decided preference for mature woodlands with sparse ground cover. Both sexes are alike. It nests on the ground among light undergrowth. The name 'warbler' is given to the bird because of its 'shivering' trill. Its food consists mainly of caterpillars, beetles, aphids and flies. It is to be found in most of Europe, apart from Scandinavia and Spain.

Blue Tit *Parus caeruleus*
The blue tit's blue cap and yellow underparts clearly distinguish it from the other tits, like the coal tit and great tit, that compete for food on bird tables.

Crested Tit *Parus cristatus*
In Britain, the crested tit is confined to the northern parts of Scotland. Its crest is very distinctive – as we see in this photograph. It is usually found in coniferous woodlands and nests in holes in old trees. This is considered to be one of John Markham's most successful photographs, and has been shown at exhibitions throughout Britain.

Long-tailed Tit *Aegithalos caudatus*
This bird measures five and a half inches, including a three-inch tail. Although very inquisitive, it is rarely seen in gardens as it prefers the woodlands and hedgerows. It has a distinctive black and white tail, with white and pink plumage on its upper and lower parts. Its nest is a magnificent piece of construction; it is dome-shaped and made of lichens, mosses and spider's webs, and lined with feathers.

Song Thrush *Turdus philomelos*
This brown-backed bird with its spotted breast can be heard singing in both town and country almost the whole of the year. It feeds upon earthworms, berries and snails and is famous for its use of an 'anvil' – a stone upon which it breaks snail shells. It nests mainly in bushes and hedges.

Linnet *Acanthis cannabina*
Most finches prefer an arboreal habitat, but the linnet, together with the twite, seem to prefer open country, allotment gardens, farm land and marshes. Linnets are fairly well distributed throughout Britain. The male, with its pink breast, is more brightly coloured than the female. Its call is very varied but melodious.

Hedgesparrow feeding Cuckoo *Prunella modularis Cuculus canorus*
The hedgesparrow or, more correctly, the dunnock is to be found in hedges, bushes and undergrowth on the edge of woodlands. The dunnock's own young have been pushed out of the nest by the young cuckoo.

Green Woodpecker *Picus viridis*
This is the largest of Britain's three woodpeckers. It is sometimes called the 'yaffle' by countrymen. It is about twelve and a half inches long and has green upper parts with a red crown and dull green underparts. The green woodpecker feeds mainly on wood-boring larvae of beetles and moths and can often be seen in large gardens probing for ants which stick to its tongue. It excavates cavities in tree trunks for nesting.

Blackbird *Turdus merula*
This picture shows a blackbird with her chicks. The female blackbird is a dark brown colour with a whitish chin and a brown bill, whereas the male has striking black plumage with a yellow bill. It is not uncommon to find albino blackbirds. They will nest almost anywhere and nowadays seem to have a marked preference for garden sheds and garages.

Starling *Sturnus vulgaris*
This bird appears to be black from a distance, but is a deep green with some purple. It changes in the winter to a closely speckled plumage. The starling is probably the most prolific bird in Britain today; in many of our cities during the winter, starling flocks numbering many thousands can be seen in the early evening returning to their warm roosts for the night. Their nests are usually made of dried grass and straw and are lined with feathers. They eat a number of pests including leatherjackets, snails and slugs.

Juvenile Bullfinch *Pyrrhula pyrrhula*
Finches have short, heavy bills, ideally suited for eating seeds. Usually the males are more brilliantly coloured than the females. This particularly applies to the bullfinches: the male is one of the most attractively coloured birds in Britain with its rose red underparts, black cap and white rump. It often visits orchards and is also very fond of honeysuckle berries.

House Sparrow *Passer domesticus*
Probably the most familiar bird in Britain, the house sparrow has become very dependent on man for its food. It can be found in almost all countries of Europe, and in parts of North Africa. The sparrow nests in holes, crevices and under eaves in buildings.

Cuckoo *Cuculus canorus*
The two most interesting characteristics of this bird are its call, which proclaims the beginning of spring, and its parasitic breeding habit of laying a single egg in another bird's nest. In Britain it appears to have a preference for meadow pipits' and dunnocks' nests. The cuckoo's egg hatches before those of its host; as the young cuckoo develops it will eject the other eggs.

Wood Pigeon *Columba oenas*
This is the largest of our pigeons, measuring sixteen inches. It can also be distinguished from other pigeons by the broad white band across the wings and a white patch on each side of its neck. In the countryside it usually roams in large flocks and is easily frightened but in the town it seems to have adapted itself to the noise and bustle. It often nests in trees in a very simple nest of twigs. In the town its nests can be seen on verandas and flower boxes.

Turtle Dove *Streptopelia turtur*
This late summer visitor to Britain is the smallest of our five breeding pigeons. It can be recognised by its slender shape and by its black tail with white edges, which is particularly conspicuous during flight. It feeds mainly on seeds and can often be seen on arable fields. It prefers open, wooded land and nests in bushes and thickets. It is not uncommon in England and Wales, but is very rarely seen in Scotland.

Bittern *Botaurus stellaris*
This large, heron-like bird is often heard in dense reed beds; but it is very seldom seen. Its deep booming call can be heard from considerable distances – up to three miles. The bittern appears to be increasing in south-east England.

Carrion Crow *Corvus corone*
The carrion crow can be distinguished from the raven by its smaller size and by its slow and regular flight – it is very rarely seen soaring. Usually solitary, it nests in trees and sometimes on sea cliffs.

Hooded Crow *Corvus corone cornix*
This sub-species of the carrion crow can be seen in Scotland and Ireland. Unmistakable from the carrion crow, with its grey back and underparts, its habits are similar to those of the carrion crow.

Great Crested Grebe *Podiceps cristatus*
This is the largest of our grebes and can be recognised — apart from its size — by its long straight neck and prominent ear tufts. However, in the winter it loses its ear tufts and it can then be confused with the red-necked grebe. Its display antics, when it appears to stand on the water, are amusing to watch.

Lapwing *Vanellus vanellus*
The lapwing is the most familiar member of the plover family in Britain. It is sometimes called the peewit because of its call *vee-veet*. It is abundant in meadows, arable fields and moorlands and can often be seen in very large flocks. It is easily recognised by its crest. When its eggs or young are in danger, the adult bird will feign injury to distract the attention of the would-be predator.

Guillemots and Kittiwakes *Uria aalge and Rissa tridactyla*
This photograph, taken on the Farne Islands, shows two of Britain's most familiar cliff-breeding birds. The guillemot can be distinguished from the black guillemot by its white breast. The latter has a very large patch of white on its wing and bright red feet.
    The kittiwake is a delightful bird to watch as it soars and glides up and down the cliffs, uttering its kittiwake call. It breeds in colonies on cliff faces, and sometimes in caves.

Puffin *Fratercula arctica*
Although similar in appearance to the penguins, the puffin is not related. It is easily distinguished from the other auks by its colourful bill, black and white plumage and orange feet. The puffin nests in holes in the ground – old rabbit or shearwater burrows are often used. It loves young sand-eels.

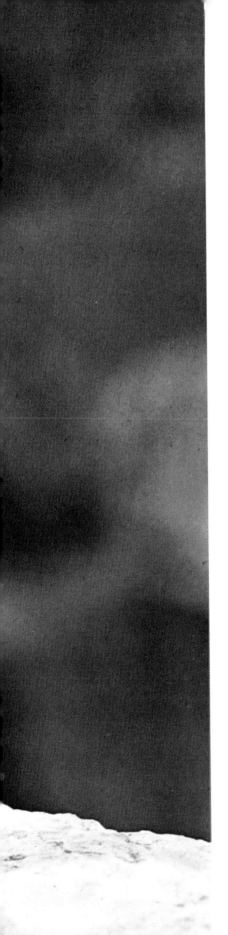

Kittiwake *Rissa tridactyla*
This is truly a bird of the sea. It has declined to take advantages of food supplies produced by man. It is very noisy during breeding seasons on northern cliffs. It has been given this name because of its call *kitti-wake*.

Canada Geese *Branta canadensis*
This handsome, introduced, black-necked goose now lives in a feral, or wild, state in Britain and is increasing considerably in numbers everywhere it finds conditions suitable. Its harsh, honking call can be heard frequently in London – especially near their flight paths between the parks.

Snowy Owl *Nyctea scandiaca*
This very attractive owl has recently returned to the Shetland Isles to breed. The adult bird is almost pure white. It frequents the Arctic tundra and barren wastes of Scandinavia and Iceland. The snowy owl is diurnal, nesting upon hummocks on the tundra.

Young Barn Owls *Tyto alba*
This most handsome of our owls can be seen — as its name implies — in or near farm buildings. It is very useful to the farmer as it lives mainly on small mammals such as mice, voles and rats. It is mainly nocturnal. Its call resembles a shriek or shrill cry rather than a hoot.

Little Owl *Athene noctua*
This is the smallest of our resident owls, measuring only nine inches. Apart from its size it can be recognised by its squat shape and large head and can often be seen during daylight. Usually it nests in holes in trees, sometimes in buildings and holes in the ground.

Tawny Owl *Strix aluco*
This is the most common of the European owls and in Britain can be seen and – more often – heard in parks and tree-lined avenues. It can be distinguished from the barn owl by its slightly larger size and its brown colouring, and from the long-eared owl by its lack of ear tufts.

Nightjar *Caprimulgus europaeus*
This is one of a nightjar series of photographs taken by John Markham, which have received acclaim wherever they have been exhibited for their technical and artistic quality. The nightjar is fairly common throughout Britain in woodlands and scrublands, but is very rarely seen due to its excellent camouflage. It is insectivorous and hunts mainly at dusk.

Stone Curlew *Burhinus oedicnemus*
Sometimes called thick-knees, the stone curlew is another bird which is more often heard than seen. Over the last decade it has declined in numbers in this country; its habitat has disappeared due to cultivation and the disappearance of the rabbit. The rabbit kept the stone curlew's habitat free of vegetation. Controlled methods are being used to maintain suitable areas for this bird to breed in Breckland.

Woodcock *Scolopax rusticola*
The woodcock can be distinguished from the snipes by its larger size, the lack of dark stripes on the crown and its more rounded wings. It is nocturnal and its almost perfect camouflage makes it very difficult to find during the day among the undergrowth of wet woodlands.

Capercaillie *Tetrao urogallus*
After disappearing in the nineteenth century, this very large bird was reintroduced into this country and its numbers are on the increase – due to the increase of coniferous woodlands in Scotland. The male can reach a length of thirty-four inches. The female is much smaller.

Greenshank *Tringa nebularia*
As its name implies, the greenshank has rather long greenish legs which distinguish it from the redshank. It breeds on moors and grasslands in northern Scotland and in Scandinavia.

Kestrel *Falco tinnunculus*
Certainly the most common of falcons in Britain, the kestrel has recently become a bird of the motorways, and can be seen hovering above the verges over the length and breadth of the country. It is becoming increasingly common in the centre of London and has nested in window boxes attached to balconies of flats. Its most distinct feature is its ability to hover when hunting for prey such as mice, small mammals and insects.

Sparrowhawk *Accipiter nisus*
This species declined in numbers in Britain during the 1960's but is recovering. It is sometimes confused with the kestrel but its colouring is different. The male has slate grey plumage on its back with a banded tail, and the female has black and brown plumage on its back.

Curlew *Numenius arquata*
This is easily distinguishable by its large and long down-curved bill. Its call is probably the best known in Britain, apart from that of the cuckoo. Its *curlee* call is often used to introduce wildlife programmes on the radio. It can be found on moors, bogs, rough fields and estuaries.

Stoat *Mustela erminea*
The stoat is coloured reddish-brown, with white to yellow on its underparts in the summer and with a prominent black tip on its tail.

Mammals

Wood Mouse *Apodemus sylvaticus*
This is also called the long-tailed field mouse. The wood mouse is well distributed over the whole of Britain and continental Europe. Its brown colouring, which varies occasionally to red, gives it an attractive appearance. The tail is the same length as the body, about three and a quarter inches.

Dormouse *Muscardinus avellanarius*
This is our only truly native dormouse, but is common only in England and Wales. It can usually be found in trees and shrubs and its food consists mainly of seeds, nuts and berries. It hibernates during the winter, usually under woodland debris. Its colouring is yellow to brown on its upperparts and creamy white on its underside.

Harvest Mouse *Micromys minutus soricinus*
This is the smallest British rodent, measuring only two and a half inches in length. It is confined mainly to East Anglia and the South of England. It feeds during the day, lives among vegetation, and is commonly found in cereal crops.

Edible Dormouse Glis glis
This is another of the photographer's famous prints which received great acclaim wherever it was displayed with the National Collection of Nature Photographs. It is a large dormouse, measuring up to seven inches. Its most remarkable feature is its beautiful bushy tail. The edible dormouse is nocturnal and lives in gardens, parks and woods.

House Mouse Mus musculus domesticus
As the name suggests, this mouse has become dependent upon man for its food and shelter. There are, however, colonies living in fields which feed on seeds. They are very restrictive in their movements, seldom straying from their homes. They can cause considerable damage to corn ricks.

Bank vole *Clethrionomys glareolus*
The bank vole is distributed over the whole of England, Wales and Lowland Scotland. Mainly vegetable feeders, these voles are good climbers and will climb trees and shrubs in search of food.

Common Shrew *Sorex araneus*
Found almost everywhere in Britain, there are many sub-species of common shrew, which vary in colour and size. They measure about three and a half inches. Like moles, they eat more than their own weight every day, so their life is one incessant hunt for food.

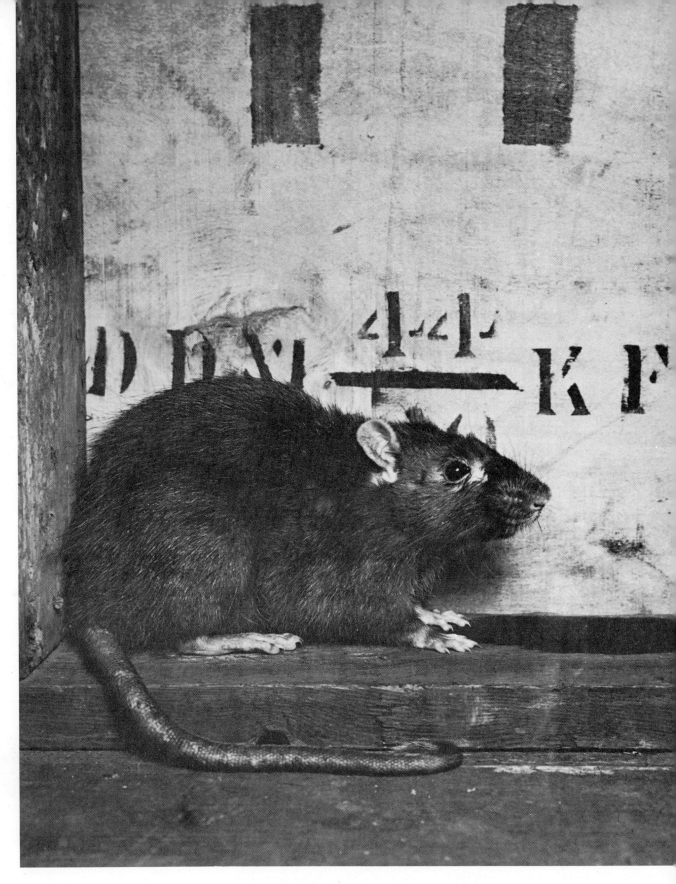

Brown Rat (black variety) *Rattus norvegicus*
These variations within the species appear quite regularly and local colonies occur. They are usually quite dark on the upperparts, unlike the true brown rat. It is sometimes referred to as the sewer rat.

Black Rat *Rattus rattus*
This rat is sometimes called the ship rat and is generally restricted to ships, ports and buildings. It is a native of Asia Minor. Completely omnivorous, it will eat anything it can digest.

Black Rat and Young *Rattus rattus*
The average litter of young is between five and eight. The young rats begin life without fur, sight or hearing. They reach the adult stage at three months.

Natterer's Bat *Myotis nattereri*
The bat is fairly common in England and Wales but is very seldom found in Scotland. It will emerge before sunset and its flight is slow and deliberate. Its food consists mainly of flies, moths and a variety of other insects and beetles.

Fox *Vulpes vulpes*
The fox usually makes its den in an old rabbit burrow or badger set. It is not very orderly with its housekeeping and the surrounds of the den usually become littered with small animal remains.

The general colour of the fox is reddish-brown, but the pattern varies considerably over the British Isles.

Fox Cub *Vulpes vulpes*
The vixen usually produces four cubs about March or April. The cubs remain in the earth for about one month and then can be seen at dusk, exercising and playing close to their earth. They reach adult size at six months.

Grey Squirrel *Sciurus carolinensis*
The grey squirrel was introduced to this country in the 19th century and has been so successful that it has now reached pest proportions, especially in parks and forests where it causes damage to young trees. It is larger than the native red squirrel and does not have ear tufts.

Red Squirrel *Sciurus vulgaris*
This is sometimes referred to as the common squirrel. It was once abundant throughout Britain but now is confined to East Anglia, northern England and Scotland. It builds its nest – or drey – in trees and feeds upon seeds, nuts, berries and sometimes bird's eggs.

Weasel *Mustela nivalis*
The weasel is the smallest of the British carnivores. measuring only a maximum of nine inches. The colour and pattern of its fur is very similar to that of the stoat. so it is best identified by its size and by its tail which does not have a black tip and is shorter – about half the length of the stoat's tail.

Stoat *Mustela erminea*
The fur of the stoat in its winter coat is ermine with black 'tabs' at the ends. Its range encompasses the whole of Britain. The stoat reaches a maximum of eleven and a half inches. A very inquisitve animal, it can often be seen standing on its hind legs watching something.

Polecat *Putorius putorius*
This is found mainly in Wales and the adjacent counties. The reduction in its numbers was probably due to excessive trapping. It is also referred to as the foul-marten because of the rather obnoxious smell it can release from its glands. The polecat has a varied diet, but feeds mainly on rats, mice, voles, rabbits and frogs.

Pine Marten *Martes martes*
Fortunately, this animal is increasing in numbers in Western Scotland due to a reduction in its persecution and an increase in woodlands. It is arboreal in its habits and expert at catching small mammals. This delightful photograph was taken at Whipsnade Zoo. The marten's beautiful, soft fur still has a commercial value in Russia and North America.

Rabbit *Oryctolagus cuniculus*
Since myxomatosis greatly reduced their numbers, rabbits' habits have changed and many now live under rocks and fallen trees, not in burrows. There is evidence that their numbers are on the increase and is causing concern among farmers.

Badger *Meles meles meles*
The badger prefers woodlands, and usually make its set in wooded or scrubby areas. It is a very good housekeeper and will replace its bedding at regular intervals. It does not hibernate during winter but does become less active. The badger keeps itself to itself: only very occasionally has it been known to attack poultry, contrary to common belief.

Young Rabbit *Oryctolagus cuniculus*

Brown Hare *Lepus europaeus occidentalis*
This is the largest of the hares found in the British Isles. Although widespread, it is not a serious pest to agriculture. Hares tend to be solitary animals. They can usually be distinguished from rabbits by their larger size, the black tips to their ears and their running, which is a series of leaps and bounds.

Hedgehog *Erinaceus europaeus*
The hedgehog is found throughout Britain and is an insectivore. As with other hibernating mammals, it builds up its fat content to sustain itself during its winter sleep. It relies upon its spines for protection, rolling into a ball when attacked by predators. Hedgehogs are quite agile climbers. Their spines usually contain fleas and mites.

Red Deer *Cervus elaphus*
This is the largest of the deer found wild in Britain. The red deer and the roe deer are our only native deer. Only the male has antlers and these are shed every year. Although originally a woodland animal, the red deer now roams the open mountain and moorland areas of Britain. The largest specimens are found in the forests of Central Europe.

Reindeer *Rangifer tarandus*
Reindeer are the only species of deer to be domesticated. Both the female and male have antlers. They live mainly upon plants – in particular mosses which grow on the tundra of Lapland. In North America they are called caribou. In the winter, reindeer migrate southwards in vast herds in search of food. Reindeer have been introduced into the Cairngorm mountains.

Fallow Deer *Dama dama*
This deer is not native to Britain but it has been here for many years and was probably introduced by the Romans. It stands about three feet in height. Its reddish-brown coat has numerous white spots during the summer but during winter it loses them. A distinctive feature is its palmate antlers.

Adder or **Viper** *Vipera berus berus*
Of the three species of snakes indigenous to Britain only the adder is venomous. It is seldom more than twenty-four inches long.

Reptiles

Grass Snake *Natrix natrix*
The grass snake reaches a length of three to four feet in this country, but on the Continent has been known to reach six feet. It feeds mainly on small frogs and toads and sometimes fish. It is an excellent swimmer.

Smooth Snake *Coronella austriaca austriaca*
This snake is the rarest of the three British species and is almost confined to the two counties of Hampshire and Dorset. It can grow to thirty inches. Its colouring is mainly a metallic grey with parallel rows of spots on its back.

Slow-worm *Anguis fragilis*
This burrowing, snake-like reptile is not a snake but a lizard which has lost its legs. In the wild it can be found in ditches, damp woodland clearings, paths, meadows and railway embankments. An adult can reach about eighteen inches and its colouring varies from grey to brown.

Eyed Lizard *Lacerta lepida*
Although this lizard is not a native of Britain, it has been included as an example of Markham's skill in photographing difficult subjects.

Sand Lizard *Lacerta agilis agilis*
Although very common in central Europe, the sand lizard is local in England. Its colour varies considerably from yellow to green depending upon its age, sex and habitat. An adult will grow to about seven inches.

Common Frog *Rana temporaria*
The common frog prefers damp, shady positions not far from ponds or streams. They are not so numerous today because of the filling-in of ponds and improved drainage systems on farms. Common frogs will travel considerable distances overland to return to their breeding site. A female can lay several thousand eggs in a season. The eggs are laid in batches quite unlike the gelatinous ropes of the toad's eggs.

Edible Frog *Rana esculenta*
The first record of the frog's introduction to Britain was in 1837 when it was brought over from France. It has since established itself in parts of Southern England. It is very similar in appearance to the common frog but can be distinguished by three light stripes on its back.

Common Toad *Bufo bufo*
The toad can be distinguished from the frog by its wrinkled and warty skin. It spends much of its life on land and, unlike the frog, is a very slow mover. It is very useful to the gardener, as it eats snails, slugs and other garden pests. The toad will climb small trees and scrub and has been known to use disused birds' nests as a home. Its eggs are laid in long gelatinous ropes.

Midwife Toad *Alytes obstetricans*
Sometimes called the bell toad the midwife toad is not a native of the British Isles, but rather of Spain, Portugal, France, Germany and Switzerland. It prefers wooded country, quite often living under the débris of fallen trees. It will also use crevices in rocks and holes in walls for sanctuary.

Black Slug *Arion ater*
Although usually black, reddish varities are often seen. It will reach a length of about seven inches.

Worms, slugs and leeches

Tree Slug *Limax marginatus*
This slug is generally well distributed throughout the British Isles, in woodlands and on rocky places and cliffs, usually hiding under rocks or in crevices. It is more active at night and during rainy periods.

Yellow Slug *Limax flavus*
This is probably the most familiar of Britain's slugs because it chooses to live near people, in garden sheds — and the garden itself. It is a scavenger, living upon refuse such as waste from the kitchen.

Earthworm *Lumbricus terrestrialis*
This is the most common of the garden earthworms. Their worm casts on lawns are very evident during autumn and winter. They feed upon organic matter pulled into their holes, sometimes sealing the entrance with débris.

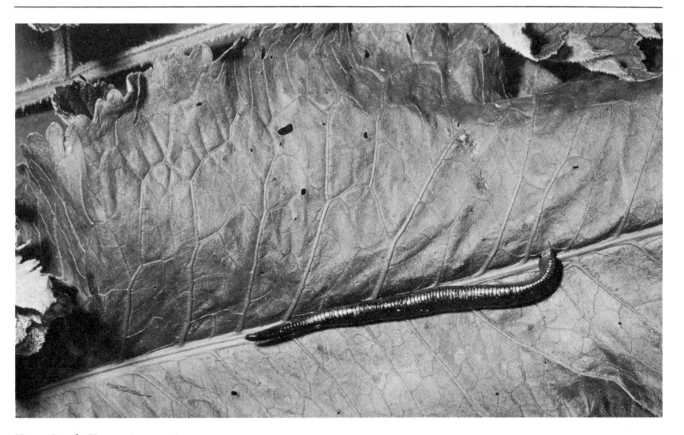

Horse Leech *Haemopis sanguisuga*
Because they were used for blood letting, these are sometimes called the medical leeches. They break down the blood for food, and can survive on one meal for about twelve months after being fully fed.

Garden Snail *Cepaea hortensis*
This snail is usually greenish-grey in colour with a yellow mantle. However, there is considerable variation in the colour of the shell. It lives in colonies in gardens and woodlands. It is a favourite food of the thrush.

Red Wasp *Vespula rufa*
This is one of the social wasps which live in huge colonies. They make their nests under the ground. Wasps do not have wax glands like the bees and therefore they are unable to make wax combs for their eggs.

Honey Bee *Apis mellifera*
This is also called the hive bee. It is not native to this country but probably originated in South-East Asia. The queen does not have wax glands like the bumble bee and she does not start a colony alone. She will collect workers to start a new colony or she will join an existing colony. A colony can exceed 50,000, the majority being worker bees.

Solitary Bee *Andrena armata*
The Andrena species are all solitary. They make their nests under very light soils and, for this reason, are sometimes referred to as miner bees. The female lays her eggs in the burrows and the progeny appear in spring.

Bumble Bee *Bombus lucorum*
The bumble bee or humble bee is a familiar sight. Bumble bees are social insects and make their nests underground. The queen bee alone survives the winter. She lays her eggs and covers them with wax which she produces from special glands. The young bees feed upon honey stored by the queen bee.

Wasp's nest *Vespula vulgaris*
This is a photograph of the common wasp. Wasps construct their nests with a sticky substance produced by a mixture of saliva and wood scrapings. The wood scrapings are obtained by the very strong jaws of the wasp, which strip the wood into fine slivers. The queen wasp deposits an egg in each cell and the subsequent larvae feed upon insects brought by the worker wasps.

Black Garden Beetle *Leistus spinibarbis*
This belongs to the group known as ground beetles. It is a very large family and there are about 350 species. Most of these beetles are small, measuring only about one inch.

Reed Beetle *Donacia semicupiea*
As its name implies, the adult reed beetle can usually be found on reeds, water lilies and other water plants. Its larvae feed on a wide variety of aquatic plants.

Silver-Fish *Lepisma saccharina*
Silver-fish are also called bristle tails. They are very common visitors to the home, especially the bathroom. Very primitive insects with long antennae and shiny scales, they are extremely difficult to catch.

Meal Worm *Tenebrio molitor*
There are about 35 different British species of meal worm. They are usually nocturnal, dark brown or black and most of the species are flightless. The meal worm lives in flour or cereals. Its larvae is a popular food of birds.

Ringlet *Aphantopus hyperanthus*
Usually to be found in shady areas such as woodland verges and glades, the ringlet has a slow lazy flight. It is common throughout Britain except for the most northern parts of Scotland. Its spot markings are variable in size and shape.

Eyed Hawk Moth *Smerinthus ocellatus*
This is one of the commoner of the British hawk moths; it is well represented in the southern part of Britain but rare in Scotland. The caterpillar has a huge appetite and gorges itself upon leaves. The eggs are laid on sallow, poplar, lime, apple, plum and pear trees.

Common Blue *Polyommatus icarus*
This butterfly is probably the most prolific of the 'blues' but is very seldom seen in large groups. Usually it is found on the chalk and limestone areas of southern England, sometimes on heathland. For food it has a preference for bird's foot trefoil. Its markings vary considerably, especially on the undersides.

Ingrailed Clay Moth *Diarsia festiva*
This moth flies in June; sometimes a second generation may be seen in August and September. It occurs over the whole of the British Isles — mainly in woods and on moorlands. The caterpillar is a pale reddish-brown. It feeds on bramble, bilberry, sallow and hawthorn.

Swallowtail *Papilio machaon*
This attractive native butterfly is also our largest. It is now confined to the Norfolk Broads. Its eggs are laid upon milk parsley (*Peucedanum palustre*), sometimes called hog's fennel, which grows in the wet meadows adjacent to the Broads.

Swallowtail *Papilio machaon*

Scorpion Fly *Panorpa germanica*
Although they look rather fearsome when examined closely, scorpion flies are quite harmless. Although not numerous, they are quite common throughout Britain.

Ladybird *Coccinella septempunctata*
These very familiar insects with their bright colours and spots are very useful to the gardener: they destroy vast quantities of green-fly and other pests. The bright colouring warns would-be predators of their bitter taste. As the latin name implies, this species has seven spots.

Crane Fly *Tipula oleracea*
Often called 'Daddy-long-legs' the crane fly is often seen when it is attracted to a light shining through a window. In the larval stage it usually lives in burrows made of mud and in shallow water and is sometimes found in decaying wood. The adult crane fly lives on nectar taken from flowers.

Ladybird *Coccinella septempunctata*
In flight.

Garden Spider *Araneus diadematus*
This spider is recognisable by the white cross on its brown back, although the brown colouring can vary considerably. Its food consists mainly of midges, and, each evening, it will eat both the prey and the web of radiating threads which it spins to catch it, before spinning a new one.

House Spider *Tegenaria domestica*
This is the smallest of this group of spiders and is common throughout Britain, mainly in houses. It spins a large, sheet-like web which can measure up to twelve inches. The spider conceals itself in a tubular nest, waiting for its victims to be caught on the web.

Goatsbeard *Tragopogon pratensis*
This plant is well distributed throughout Britain. The flowers open only during the daytime.

Plants

Columbine *Aquilegia vulgaris*
This very common garden plant can be found in the wild in woods, hedgerows and ditches. It is usually a blue-purple colour but in the garden it can be white, pink, yellow or red.

Honeysuckle berries *Lonicera periclymenum*
This very sweet-scented climbing plant is fairly widespread in woodlands and hedgerows.

Red Campion *Silene dioica*
This scentless, hairy plant can be found in woods — usually on well-drained soils and sometimes on cliffs, particularly if the cliffs have been used by nesting birds.

Leaves of the Horse Chestnut
Aesculus hippocastanum
These are the leaves of a very large deciduous tree which was introduced to this country from south-eastern Europe and can be found in many parks in Britain. It is very popular with young boys for its 'conkers'.

Pansy or Heartsease *Viola tricolor*
This is found mainly on cultivated ground, but sometimes on grasslands. Its colour and patterns are very variable.

Wild Dog Rose 'hips' *Rosa canina*
'Hips' is the name given to the fruits of the wild rose. It is the most common of our wild roses and is widespread on the verges of woods, hedgerows and thickets.

Summer Snowflake *Leucojum aestivum*
This plant is also known as the Loddon lily. As a native plant it occurs very locally in wet meadows. It is also cultivated and escapes from gardens to grow wild.

Rough Meadow Grass *Poa trivialis*
Rough meadow grass is found mainly in meadows and open areas and it is common throughout Britain.

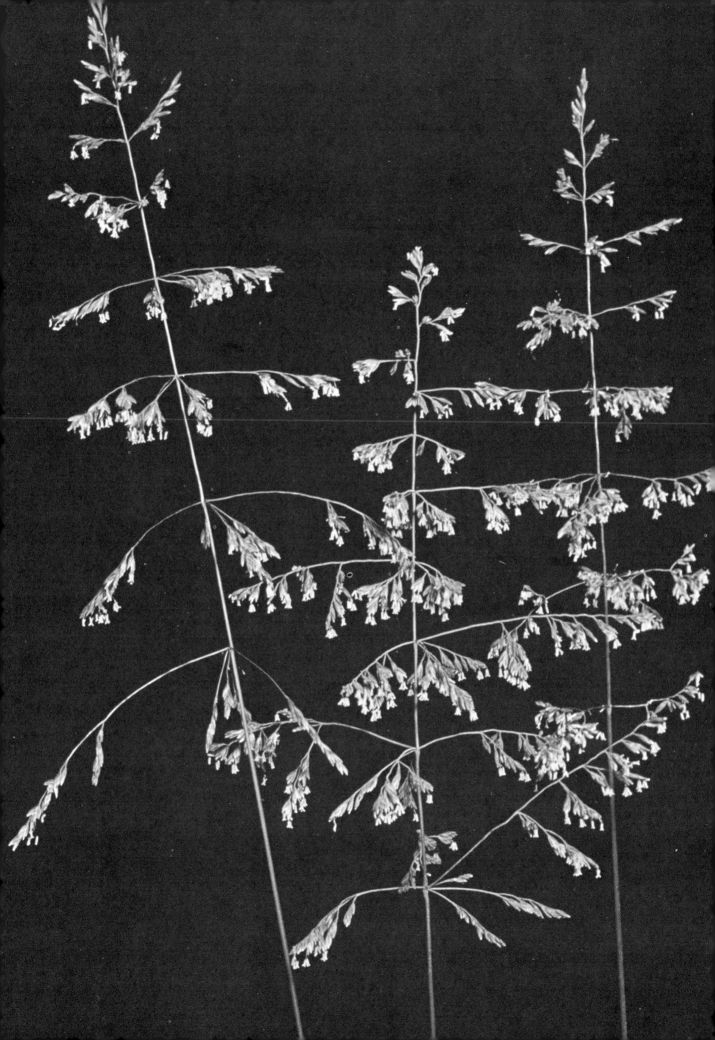

Grape Hyacinth *Muscari atlanticum*
This attractively shaped and coloured plant is to be found locally in Norfolk and Suffolk and – rather more rarely – in the Cotswolds. It is a different species to the one commonly grown in gardens.

Dandelion *Taraxacum officinale*
When seen without prejudice, this weed can produce a beautiful carpet of yellow in our meadows and verges during the months of May and June.

Ox-eye Daisy *Chrysanthemum leucanthemum*
This large daisy, sometimes called ox-eye daisy or marguerite is found throughout Britain on railway embankments and other grassy places.

Gentian *Gentiana septemfida*
This is not truly a British species but is becoming common in gardens; this is an escape near to the photographer's garden.

Wall Pennywort *Umbilicus rupestris*
This unusually shaped plant with its disc-like, fleshy leaves is common on walls in Wales and in the West Country.

Foxglove *Digitalis purpurea*
This attractive, biennial plant is widespread, being common in woods, roadside verges, heaths and mountain rocks. It usually grows on acid soils. The drug digitalin is extracted from this plant and used in treating some heart diseases.

Royal Fern *Osmunda regalis*
This handsome British fern will grow to over six feet in favourable conditions. It prefers fens, wet heaths and dampish woodlands.

Wild Daffodil *Narcissus pseudonarcissus*
In the wild this well-known plant has much smaller flowers than the garden variety. It is found locally in damp woodlands and meadows throughout England and Wales.

Spanish Bluebell *Endymion hispanicus*
As its name implies, this flower is native of Spain; it is also found in Portugal. It is very common in gardens, from which it has escaped and naturalised in a few places in Britain.

Dwarf Cornel *Chamaepericlymenum suecicum*
This attractive little plant with its blackish flower heads with four white bracts looking like petals, is often hidden by bilberries or heather in the Highlands of Scotland. It is very rare in England.

Primrose *Primula vulgaris*
The primrose is widespread throughout Britain and is easily recognisable by its yellow flowers and crinkly leaves. It prefers banks, woodland glades and scrubby areas.

Hemp Agrimony *Eupatorium cannabinum*
A common perennial plant of the fens and marshes, hemp agrimony is found in wet woodlands throughout Britain, although less common in Scotland.

Creeping Buttercup *Ranunculus repens*
The creeping buttercup is widespread in wet meadows, pastures and woods – especially on heavy soils.

Chickweed Wintergreen *Trientalis europaea*
This slender little perennial plant is native to northern pinewoods and grows in mossy, grassy places. It grows to about four inches high.

Lily of the Valley *Convallaria majalis*
This fragrant common garden flower is native to dry, calcareous woodlands and is widespread, although local, throughout England and parts of Wales. It is rarer in Scotland.

Lords and Ladies *Arum maculatum*
This common plant grows in woodlands and shady hedges throughout the British Isles except for northern Scotland. After the flower's hood withers, a spike of green, later bright red, berries develop, and these fruits are poisonous. In the sixteenth century, the tubers of this plant were used to starch courtier's ruffs. It is also called cuckoo-pint.

Common Oak Catkins *Quercus robur*
This is probably the most familiar tree in Britain, with its wide crown and large brownish-grey fissured trunk. It is found throughout Britain but does not thrive on acid peat or shallow limestone.

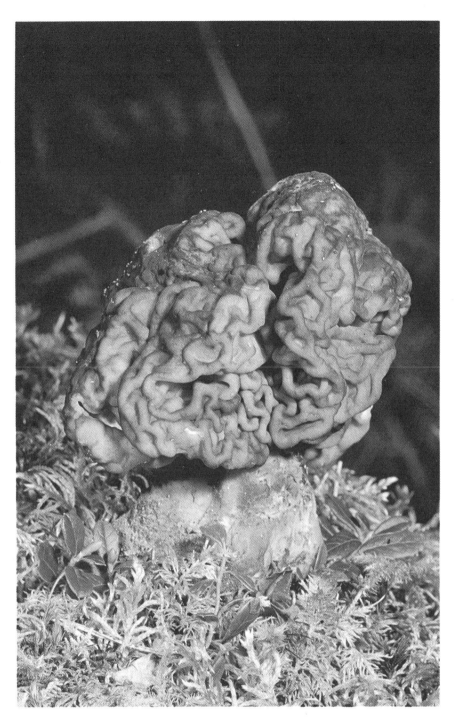

Fungus *Gyromitra esculenta*
This is an uncommon fungus found in the sandy soils of coniferous woodlands in Scotland. It is poisonous.

Bugle *Ajuga reptans*
This is a common plant found in damp woodlands and shady grassy areas throughout the British Isles. The flowers are blue, rarely pink or white. It grows to about six inches.

Pyramidal Orchid *Anacamptis pyramidalis*
This orchid is fairly common on chalk or limestone pasture land and sand-dunes and is easily recognised by the pyramidal shape of the pink flower spike. Often white flower heads are found. The pyramidal shape is more obvious during the earlier stages of the development of the flower head. It has a wide distribution throughout Europe, but is not found in northern Scandinavia.

Marsh Helleborine *Epipactis palustris*
This is the most attractive of the helleborines found in Britain, the flowers being more attractively coloured and larger than any other helleborine. Another unusual feature of this plant is its vegetative method of growth. Underground stems grow upwards each year to form a flowering shoot. It can be found in fens and dune slacks in England, Wales and southern Scotland.

Common Butterwort *Pinguicula vulgaris*
This plant is so named because of the pale yellow colour of its basal rosette of leaves. The leaves are covered with small hairs and sticky glands for trapping insects which it digests. Usually it is found on boggy areas, at the sides of lochs and on wet heathlands. It is very rare in the South, but widespread elsewhere on acid soils.

Hart's Tongue Fern *Phyllitis scolpendrium*
This popular fern, that grows in many gardens, can be seen in the wild on rocks, walls and damp areas, within open woodlands and mainly in the west. It is the only true fern in Britain without divided leaves. The spore cases are attached to the backs of the leaves.

Alkanet *Pentaglottis sempervirens*
Alkanet flowers are similar to the forget-me-not but the leaves are much broader. It has become naturalised in hedgerows, on earth banks and in borders of woods. It flowers from April onwards. It is sometimes referred to as the green alkanet.

Yew *Taxus baccata*

The yew is indigenous to Britain. Although usually seen along parish boundaries and churchyards, there are two very fine woods at Kingley Vale, Sussex and Butser Hill, Hampshire. The yew is a very long-lived tree and some of the specimens at Kingley Vale are considered to be over 500 years old. The yew will grow on different soils but seems to prosper most on limestone and chalk. In the past the wood was used for making bows. Yews also played a role in religious ceremonies conducted by the Druids.

Index

Acanthis cannabina, 32
Accipiter nisus, 55
adder, 78
Aegithalos caudatus, 31
Aesculus hippocastanum, 105
agrimony, hemp, 114
Ajuga reptans, 119
albino blackbird, 34
Alcedo atthis, 21
alkanet, 121
Alytes obstetricans, 86
Anacamptis pyramidalis, 119
Andrena armata, 92
Anglesey, 10
Anguis fragilis, 80
'anvil', 31
Aphantopus hyperanthus, 96
Apis mellifera, 92
Apodemus sylvaticus, 58
Apus apus, 26
Aquilegia vulgaris, 104
Araneus diadematus, 102
Arion ater, 86
Arum maculatum, 116
Athene noctua, 50
auk, 45

badger, 73
bankvole, 61
barn owl, 49, 50
bat, natterer's, 64
Beachy Head, 14
bee,
    bumble, 93
    honey, 92
    solitary, 92
beetle,
    black garden, 96
    ground, 96
    reed, 94
bell toad, see midwife toad
Bird Act 1967, 16
bittern, 40
black garden beetle, 96
black guillemot, 45
black rat (ship rat), 63
blackbird, 34
    albino, 34
blackcap, 25
bluebell, 14
    Spanish, 113
blue tit, 19, 20, 26, 28
Bombus locorum, 93
Botaurus stellaris, 40
Branta canadensis, 47
Breckland, 52
bristletail, see silver-fish
Brooke Bond Tea picture card
    album, 13
brown hare, 74
brown rat (sewer rat), 62
Bufo bufo, 86
bugle, 119
bullfinch, 36

bumble bee, 93
Burhinus oedicnemus, 52
Buster Hill, 122
buttercup, creeping, 114
butterfly, 98, 99
    common blue, 98
    swallow-tail, 99, 100
butterwort, common, 120

Cairngorm mountains, 10, 77
campion, red, 104
Canada geese, 47
capercaillie, 53
Caprimulgus europaeus, 50
Carduelis cannabina, 17
caribou, see reindeer
carrion crow, 40, 41
catkins, 116
Cervus elaphus, 76
chaffinch, 10, 18
Chamaepericlymenum suecicum, 113
chestnut, horse, 105
chickweed wintergreen, 115
Chrysanthemum leucanthemum, 109
clay moth, ingrailed, 99
Clethrionomys glareolus, 61
coal tit, 19, 28
Coccinella septempunctata, 100, 101
Columba oenas, 39
columbine, 104
Committee on Photography, 16
common blue, 98
common butterwort, 120
common frog, 84
common shrew, 61
common squirrel, see red
    squirrel
common toad, 86
common wasp, 90, 93
'conker', 105
Convallaria majalis, 115
cornel, dwarf, 112
Coronella austriaca austriaca, 80
Corbus corone, 40
Corvus corone cornix, 41
Country Life, 13
crane fly, 101
creeping buttercup, 114
crested tit, 28
crow,
    carrion, 40, 41
    hooded, 41
cuckoo, 33, 38, 56
cuckoo pint, see lords and
    ladies
Cuculus canorus, 33, 38
curlew, 56
    stone, 10, 52

'daddy-long-legs, see crane fly
daffodil, wild, 113
daisy, ox-eye, 109
Dama dama, 78
dandelion, 108
deer, 76, 77, 78
    fallow, 78
    red, 76
    roe, 76
Devon, 10
Diarsia festiva, 99, 100

digitalin, 111
Digitalis purpurea, 111
diver, red-throated, 13, 14
dog daisy, see ox-eye daisy
Donacia semicupiea, 94
dormouse, 58
    edible, 60
dotterel, 10, 11, 13
dove, turtle, 39
dunnock, see hedgesparrow
dwarf cornel, 113

earthworm, 89
East Anglia, 59
edible dormouse, 60
edible frog, 85
Endymion hispanicus, 113
Epipactis palustris, 119
Erinaceus europaeus, 75
ermine, 69
Eupatorium cannabium, 114
eyed hawk moth, 96
eyed lizard, 83

Falco tinnunculus, 54
falcon, 54
    kestrel,
fallow deer, 78
Farne Islands, 45
fern, 111, 120
    Hart's tongue, 120
    royal, 111
Ficedula hypoleuca, 18
Field, The, 13
*Field Annual on Nature
    Photography, The*, 16
finch, 18, 36
    bull-, 36
    chaf-, 10, 18
    linnet, 32
fly,
    crane, 101
    scorpion, 100
flycatcher, pied, 18
foul-marten, see polecat
fox, 65
    cub, 66
foxgloves, 111
Fratercula arctica, 45
Fringilla coelebs, 18
frog, 86
    common, 84, 85
    edible, 85

garden snail, 90
garden spider, 102
garden warbler, 11, 25
Glis glis, 60
goatsbeard, 102
goose, 47
    Canada, 47
grape hyacinth, 108
grass, rough meadow, 106
grass snake, 80
grasshopper warbler, 10
Great Northern Cemetery, 13
great tit, 20, 25, 28
grebe, 42
    great crested, 42
    red-necked, 42
green alkanet, see alkanet
greenshank, 54

green woodpecker, 34
grey squirrel, 66
ground beetle, 96
guillemot, 45
    black, 45
Gun & Game, 13

Haemopis sanguisuga, 89
hare, 74
    brown, 74
hart's tongue fern, 120
harvest mouse, 59
heartsease, see pansy
hedgehog, 75
hedgesparrow (dunnock), 33
helleborine, 119
    marsh, 119
hemp agrimony, 114
hide, 13
Higham, Walter E., 10
Hillside Rock and Dale (Pike), 12
hive bee, see honey bee
hog's fennel, see milk parsley
honey bee, 92
honeysuckle, 104
hooded crow, 41
horse chestnut, 105
horse leech, 89
Hosking, Eric, F.R.P.S., 12, 13, 13–14, 14, 16
house mouse, 60
house sparrow, 36
house spider, 102
humble bee, see bumble bee
hyacinth, grape, 108

Illustrated London News, 13

Kearton brothers, 10
kestrel, 54, 55
kingfisher, 21
Kingley Vale, 122
kittiwake, 45, 46

ladybird, 100, 101
Lacerta agilis agilis, 83
Lacerta lepida, 83
lapwing (peewit), 13, 43
leech, horse, 89
Leistus spinibarbis, 96
Lepisma saccharina, 95
Lepus europaeus occidentalis, 74
Leucojum aestivum, 106
Lily of the valley, 115
Limax flavus, 86
Limax marginatus, 88
linnet, 17, 32
little owl, 50
lizard
    eyed, 80, 83
    sand, 83
    slow-worm, 80
loddon lily, see summer
    snowflake
London Calling, 13
long-tailed field mouse, see
    wood mouse
long-tailed tit, 31
Lonicera periclumenum, 104
lords and ladies, 116
Lowes, Harold, 14
Lumbricus terrestrial ris, 89
Luscinia megarhynchos, 21

Mammal Society for Great Britain, 14
marguerite, *see* ox-eye daisy
marsh helleborine, 119
marsh tit, 20
marten, pine, 70
Martes martes, 70
martin, sand, 22
Masterpieces of Bird Photography, 13
meadow grass, rough, 106
meal worm, 95
'medical leech', *see* horse leech
Melandrium dioicum, 104
Meles meles meles, 73
Micromys minutus soricinus, 59
midwife toad, 86
milk parsley (hog's fennel), 99
miner bee, *see* solitary bee
mole, 61
Motacilla alba, 23
moth,
   eyed hawk, 96
   hawk, 96
   ingrailed clay, 99
mouse, 14
   harvest, 59
   house, 60
   wood, 15, 58
Muscardinus avellanarius, 58
Muscari racemosum, 108
Mustela erminea, 56, 69
Mustela nivalis, 68
Myotis nattereri, 64
myxomatosis, 73

Narcissus peudonarcissus, 113
National Collection of Nature Photographs, 14, 60
Natrix natrix belvetica, 80
natterer's bat, 64
Nature Conservancy's Advisory Committee on Photography, 14
Nature Photographic Society, 14
New Naturalist Series, 10, 13
nightingale, 21
nightjar, 10, 50
Norfolk, 10, 16
Norfolk Broads, 99
Numenius arquata, 56
Nyctea scandiaca, 49

oak, 116
   catkins, 116
orchid, pyramidal, 119
Osmunda regalis, 111
owl, 49, 50
   barn, 49, 50
   little, 50
   snowy, 49
   tawny, 50
ox-eye daisy, 109

Panorpa germanica, 100
pansy (heartsease), 105
Papilio machaon, 99, 100
parsley, milk, 99
Parus ater, 19
Parus caeruleus, 26, 28
Parus cristatus, 28
Parus major, 25

Parus palustris, 20
Passer domesticus, 36
peewit, *see* lapwing
penguin, 45
pennywort, wall, 110
Pentaglottis sempervirens, 121
Photographic Journal, The, 12
Phyllitis scolpendrium, 120
Phylloscopus sibilatrix, 27
Picus viridis, 34
pied flycatcher, 18
pied wagtail, 23
pigeon, 39
   wood, 39
Pike, Oliver, 10, 12
pine marten, 70
Pinguicula vulgaris, 120
plover, 43
Poa trivialis, 106
Podiceps cristatus, 42
polecat, 70
Polyommatus icarus, 98
primrose, 113
Primula vulgaris, 113
Pruneall modularis, 33
puffin, 45
Putorius putorius, 70
pyramidal orchid, 119
Pyrrhula pyrrhula, 36

Quercus robur, 116

rabbit, 52, 73, 74
Rana esculenta, 85
Rana temporaria, 84
Rangifer tarandus, 77
Ranunculus repens, 114
rat,
   black (ship), 63
   brown (sewer), 62
   true brown, 62
Rattus norvegicus, 62
Rattus rattus, 63
raven, 40
red campion, 104
red deer, 76
red-necked grebe, 42
redshank, 54
red squirrel, 66, 67
red-throated diver, 13, 14
red wasp, 90
reed beetle, 94
reindeer, 77
ringlet, 96
Riparia riparia, 22
Rissa tridactyla, 45, 46
robin, 12, 13
rodent, 59
roe deer, 76
Rosa canina, 105
rose hip, wild, 105
rough meadow grass, 106
royal fern, 111
Royal Photographic Society, 12, 14
   Exhibitions Committee, 14
   Fellowship and Associateship Admission Committees (Nature) Panel, 14
Royal Society, 13
Royal Society for the Protection of Birds, 14

sand lizard, 83
sand martin, 22
Sciurus carolinensis, 66
Sciurus vulgaris, 67
Scolopax rusticola, 52
scorpion fly, 100
Shetland Isles, 49
ship rat, *see* black rat
shrew, common, 61
silver-fish, 95
slow-worm, 80
slug, 86, 88
   black, 86
   tree, 88
   yellow, 86
Smerinthus ocellatus, 96
smooth snake, 80
snail, garden, 90
snake
   adder, 78
   grass, 80
   smooth, 80
snipe, 52
snowy owl, 49
solitary bee, 92
song thrush, 31
Sorex araneus, 61
Spanish bluebell, 113
sparrow, house, 36
sparrowhawk, 55
Sphere, The, 13
spider, 14
   garden, 102
   house, 102
squirrel
   grey, 66
   red, 66, 67
starling, 12, 35
stoat, 56, 68, 69
stone curlew (thick-knees), 10, 52
Streptopelia turtur, 39
Strix aluco, 50
Sturnus vulgaris, 35
summer snowflake, 106
swallow, 22, 26
swallow-tail, 99, 100
swift, 26
Sylvis borin, 25

Taraxacum officinale, 108
Tardus merula, 34
tawny owl, 50
Taxus baccata, 122
Tegenaria domestica, 102
Tenebrio molitor, 95
Tetrao urogallus, 53
thick-knees, *see* stone curlew
thrush, song, 31
Tipula oleracea, 101
tit, 25
   blue, 19, 20, 26, 28
   coal, 19, 28
   crested, 28
   great, 20, 25, 28
   marsh, 20
   long-tailed, 31
   willow, 10, 11, 20
toad, 84
   common, 86
   midwife, 86
Tragopogon pratensis, 102
tree slug, 88

Trientalis europaea, 115
Tringa nebularia, 54
true brown rat, 62
Turdus philomelos, 31
turtle dove, 39
twite, 17, 32
Tyto alba, 49

Umbilicus rupestris, 110
Uria aalge, 45

Vanellus vanellus, 43
Vespula rufa, 90
Vespula vulgaris, 93
Viola tricolor, 105
viper, *see* adder
Vipera berus berus, 78
vole, bank, 61
Vulpes vulpes, 65, 66

wagtail, pied, 23
wall pennywort, 110
warbler, 25
   garden, 11, 25
   grasshopper, 10
   wood, 10, 27
wasp, 90, 93
   common (nest), 93
   red, 90
weasel, 68
wild daffodil, 113
wild rose hip, 105
willow tit, 10, 11, 20
woodcock, 10, 13, 52
wood mouse (long-tailed field mouse), 15, 58
woodpecker, green, 34
wood pigeon, 39
wood warbler, 10, 27
worm, meal, 95

'yaffle', *see* green woodpecker
yew, 122

Zoological Photographic Club, 14
Zoological Society, 12